The Oxford Piano Method

MIXED DOUBLES
More Duets with a Difference

Pauline Hall

Music Department
OXFORD UNIVERSITY PRESS
Oxford and New York

Acknowledgements

Illustrations by John Taylor

The following copyright material is included:

'Courante' and 'Galop' from *A Seven Piece Suite* by Elsie Wells. © 1979 Oxford University Press.

'Gavotte' by Samuel Wesley, arr. W. Appleby. Reproduced by permission of Banks Music Publications.

'Graceful Dance' from *Progressive Duets Book 1* by Adam Carse. Reproduced by permission of Stainer & Bell Ltd., London, and Galaxy Music Corporation, Boston.

'Missa Ramgoat' from *Twelve Jamaican Duets* by Barbara Kirby-Mason. © 1970 Oxford University Press.

'Rumba' from *Ten Rhythmical Dances* by Gerard Hengeveld. Reproduced by permission of Broekmans & Van Poppel B.V., Amsterdam.

'Russian Dance' and 'Spanish Dance' from *A Suite of Five Easy Duets* by T. A. Johnson. Reproduced by permission of Alfred Lengnick & Co. Ltd.

CONTENTS

Missa Ramgoat

Jamaican folk-song
arr. Barbara Kirby Mason

Cheerfully

Missa Ramgoat

Jamaican folk-song
arr. Barbara Kirby Mason

If you think this piece is too short, you could repeat it.

Courante

Elsie Wells

Courante

Elsie Wells

Russian Dance

Thomas Arnold Johnson

Russian Dance

Thomas Arnold Johnson

'Presto' means fast – but *please* not until you can play it slowly!

Graceful Dance

Adam Carse
(1878–1958)

Graceful Dance

Adam Carse
(1878–1958)

Secondo

Polonaise

Antonio Diabelli
(1781–1858)

Allegretto

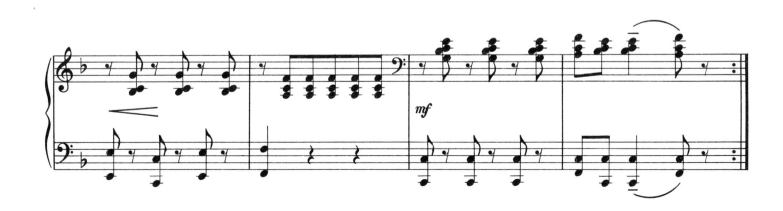

Polonaise

Antonio Diabelli
(1781–1858)

Pilgrims' March

Felix Mendelssohn
(1809–1847)
arr. Pauline Hall

Pilgrims' March

This comes from Mendelssohn's 'Italian' Symphony. You'll enjoy
listening to the full orchestra playing this, so look out for it.

Felix Mendelssohn
(1809–1847)
arr. Pauline Hall

Secondo

All Through the Night

Welsh folk-song
arr. Pauline Hall

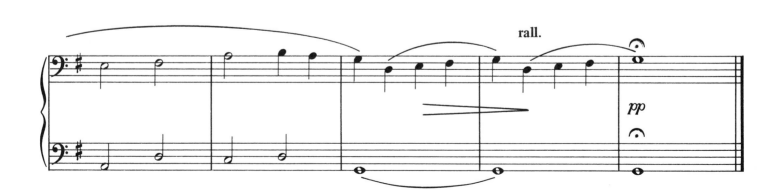

All Through the Night

Welsh folk-song
arr. Pauline Hall

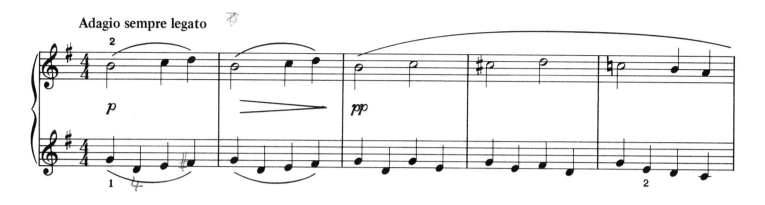

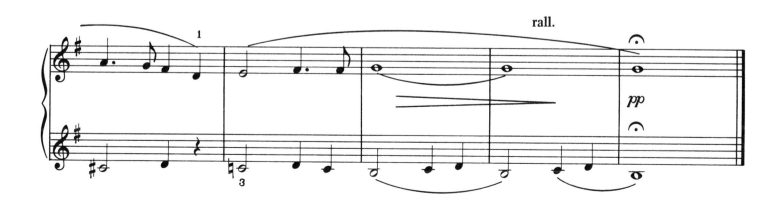

A listening game: Snap!

1st player plays a note – any one will do, but somewhere in the middle two octaves of the piano is best. He plays his note several times, listening to it carefully to fix it in his mind. He then turns his back on the piano so that he can't see the keyboard.

2nd player (who knows the chosen note) slowly plays the surrounding notes using one finger, and somewhere amongst them includes the 1st player's note.

If the 1st player recognizes it, he shouts 'Snap!' and wins a point. If he misses it, or chooses the wrong one, the 2nd player scores a point.

Then the players change places.

Here is an example:

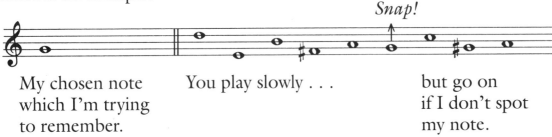

My chosen note   You play slowly . . .   but go on
which I'm trying   if I don't spot
to remember.   my note.

Evening bells

Pauline Hall

In this duet, the top part can be played by someone who hasn't learnt the piano. They use 2 fingers, and you must show them where their two notes are – 'A' in each hand. You must explain that there are 4 beats in each bar, and that they play on the 2nd and 3rd beats (except in bar 7, where they don't play at all). The bars are numbered to make it easier.

You *must* practise your part on your own, so that you can play it without mistakes.

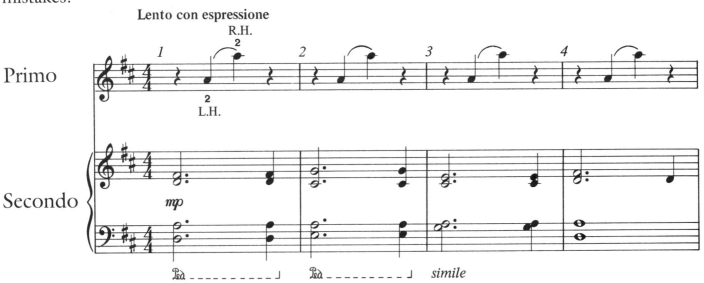

Spanish Dance

Thomas Arnold Johnson

(1st time continue overpage)
Fine

Spanish Dance

Thomas Arnold Johnson

(1st time continue overpage)

D.C. al Fine

Casey Jones

American railroad ballad
arr. Pauline Hall

Can you get the effect of the train (complete with its whistle)? Keep it quiet
until you have the tune in bar 13, when it is Primo's turn to keep quiet.

Casey Jones

American railroad ballad
arr. Pauline Hall

Here's a little memory test. To turn the page, you're going to have to memorize
the first bar that you play. Take a good look at it before you let the Secondo
start!

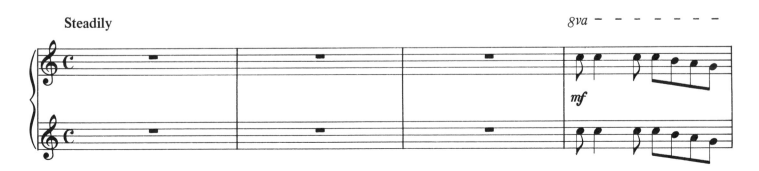

Secondo

Secondo

Gavotte

Samuel Wesley
(1766–1837)
arr. William Appleby

Moderato

Gavotte

Samuel Wesley
(1766–1837)
arr. William Appleby

Although the notes in the 7th line are divided between the hands,
really it's just one long melody.

Ten green bottles

arr. Pauline Hall

You will notice that in this piece the parts are set out above each other and not on separate pages. This means that you can see the other player's part more easily, but you have to be careful to jump to your own part at the end of each line.

Another listening game: Pool gazing

When you look in a rock-pool on the sea-shore, you first of all see the reflection of the sky on the surface of the water. But if you try looking down *beneath* the surface, you can see things deeper down in the water – perhaps a starfish or a little crab.

Think of listening to two notes in this way. The top one is on the surface, and easy to hear. The lower one is deeper down, so listen beneath the surface, and you'll hear it.

In this game you can choose which note you are going to sing – upper or lower. The surface note scores one point if you get it right; the deeper note scores five points.

Ready?
1st player plays one of the examples below.
2nd player sings his or her chosen note (no points if it's wrong!).
Change places. The best score out of ten turns wins.

Good luck, and good fishing!

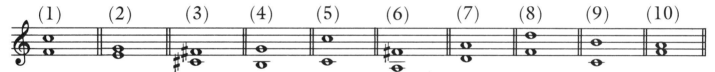

Andante

Antonio Diabelli
(1781–1858)

Andante

Antonio Diabelli
(1781–1858)

Galop

Elsie Wells

leggiero – lightly

Galop

Elsie Wells

If you are wise, you will play this quite steadily until your fingers know exactly where to go.

Norwegian Dance No.2

Edvard Grieg
(1843–1907)
arr. Pauline Hall

Allegretto tranquillo e grazioso

Norwegian Dance No.2

Edvard Grieg
(1843–1907)
arr. Pauline Hall

Secondo

Hunsdon House

Old English dance tune
arr. Pauline Hall

Hunsdon House

Old English dance tune
arr. Pauline Hall

Look out for the rhythm ♩ ♪♪♩

and be careful that it doesn't turn into ♩ ♪♩ ♪ !

Rumba

It is a help to count this typical rumba rhythm in quavers, grouped like this:
1 2 3, 1 2 3, 1 2. But watch out for where this changes in the 2nd, 3rd, and last
line of page 44. Be careful not to be distracted by the rhythm in the top part.

Gerard Hengeveld

Rumba

Primo: Your hands stay in the same position over these notes:

Gerard Hengeveld

f (repeat p)

Secondo